FAÇADE- SHOP

店面

《新空间》编辑组 编
张晨 译

辽宁科学技术出版社

FAÇADE- SHOP
店面

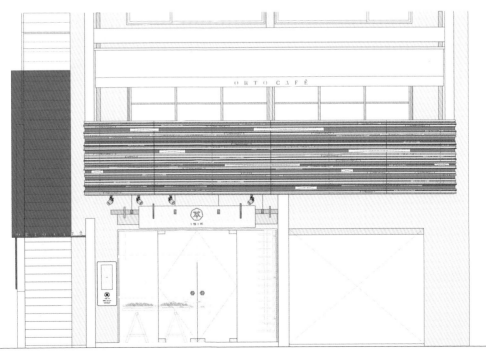

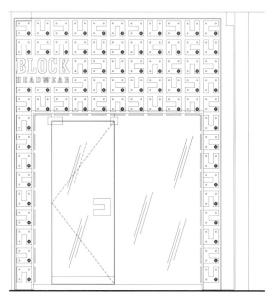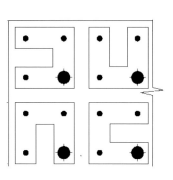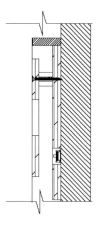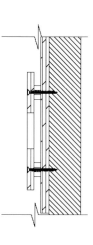

IDEAL BLANK

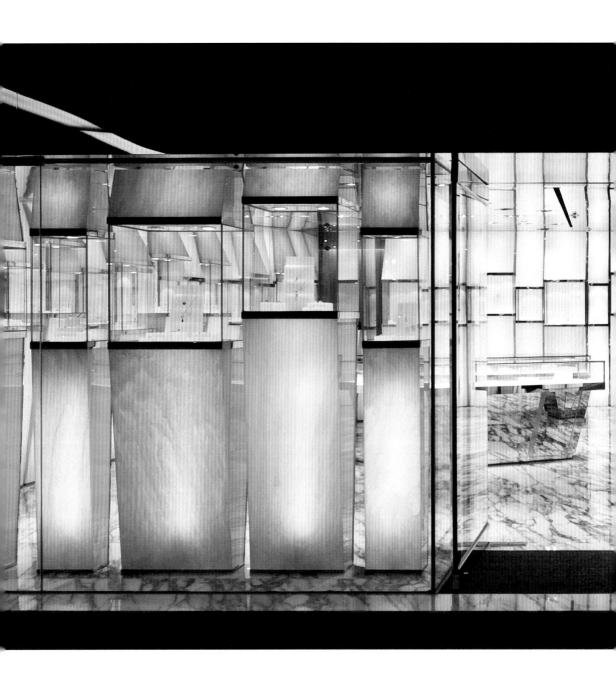

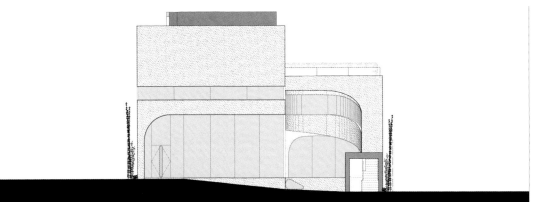

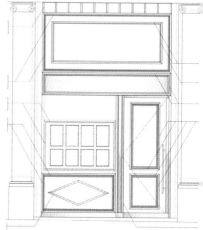

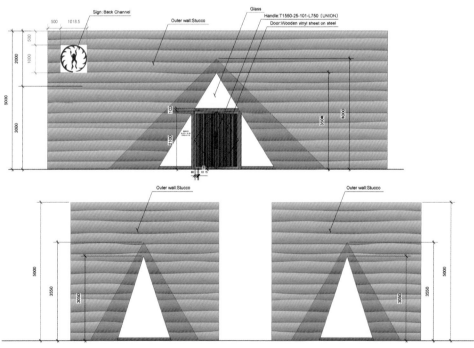

029

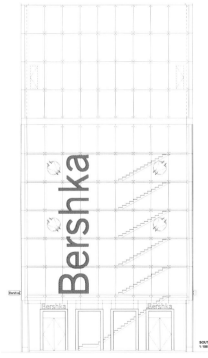

SOUTH ELEVATION
1:100

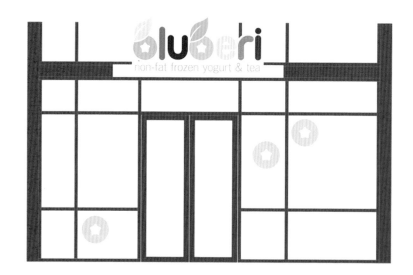

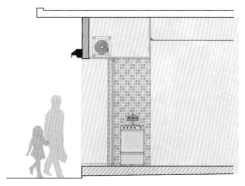

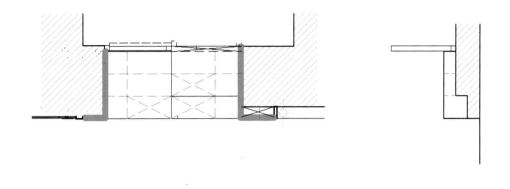

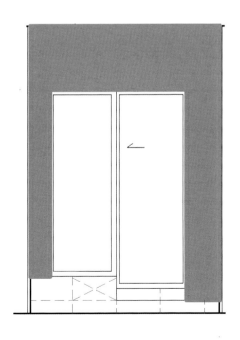

043

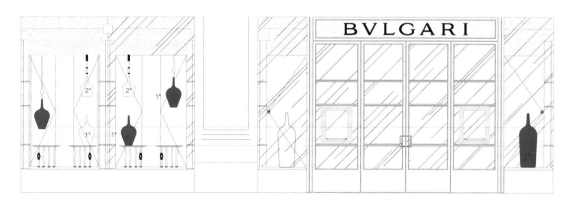

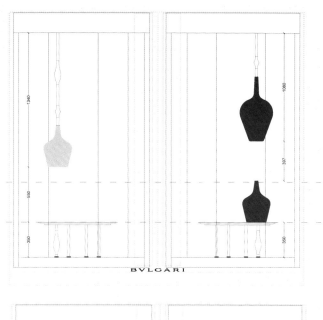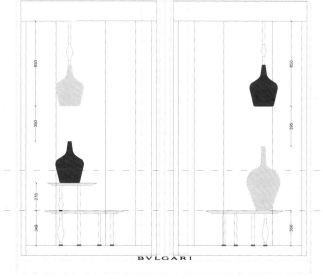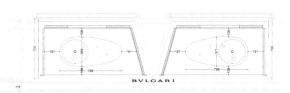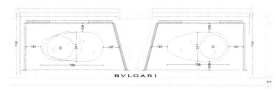

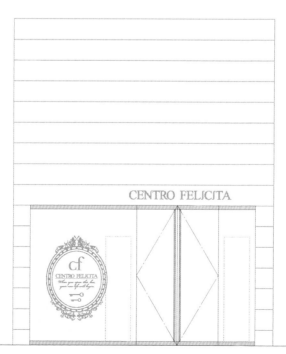

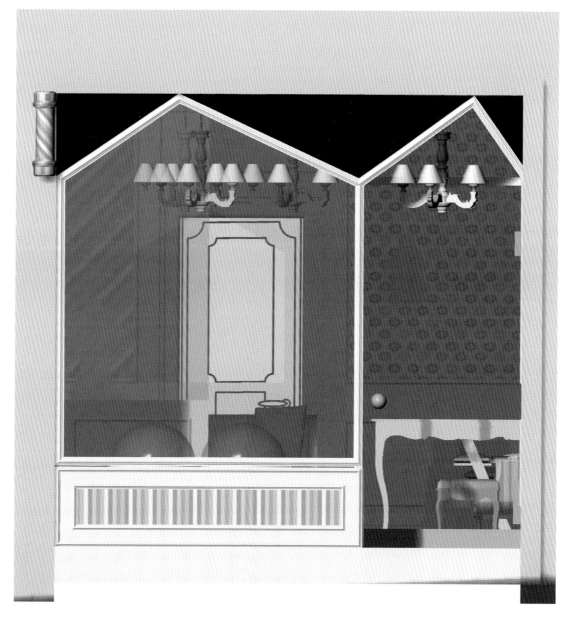

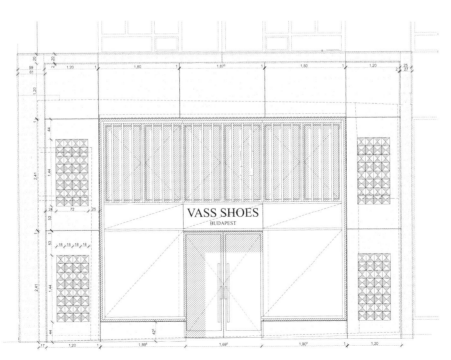

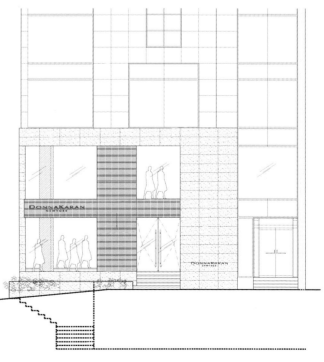

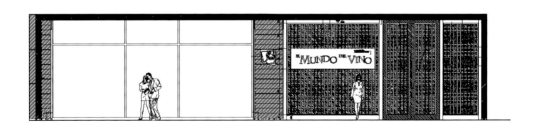
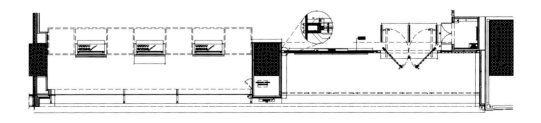

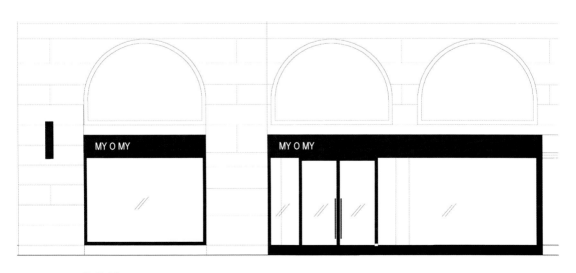

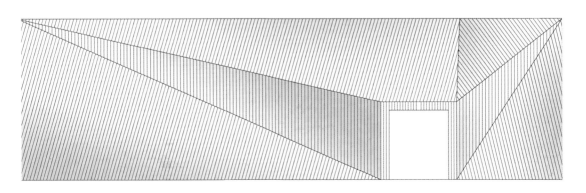

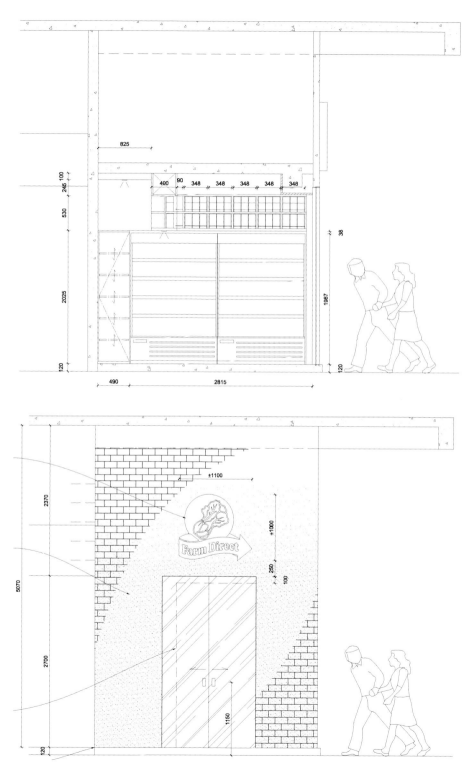

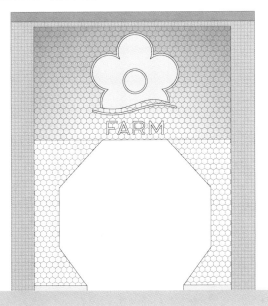

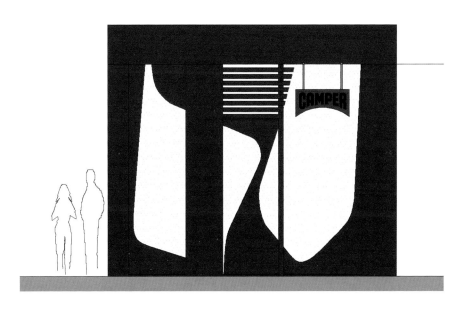

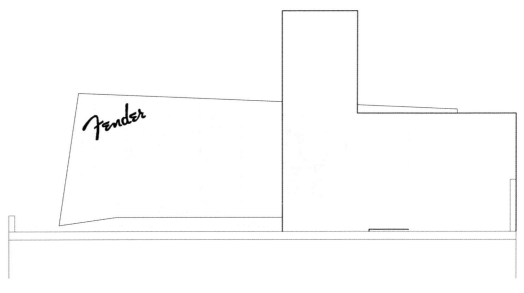

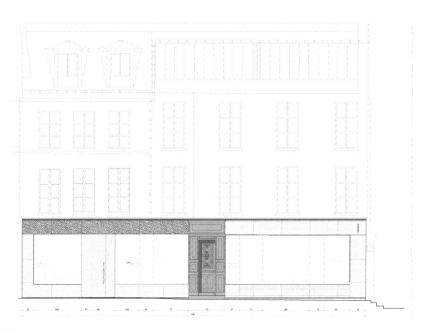

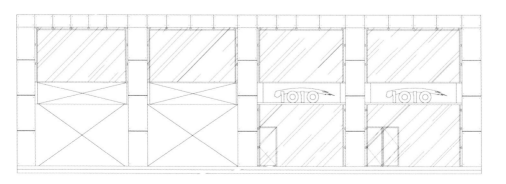 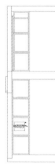

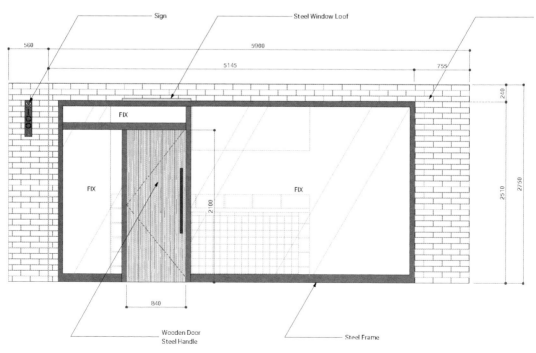

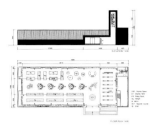
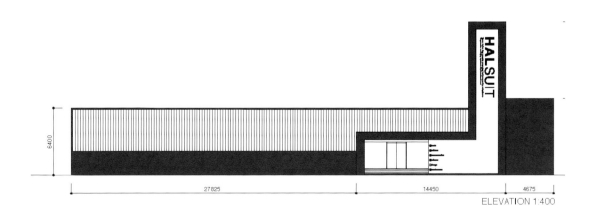

ELEVATION 1:400

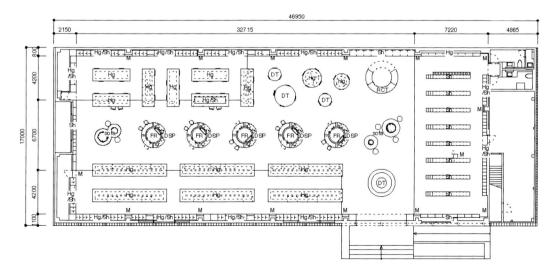

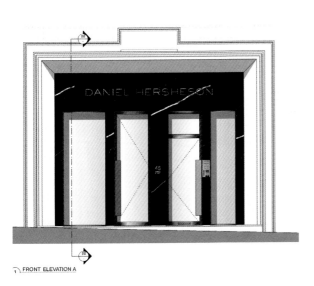

FRONT ELEVATION A

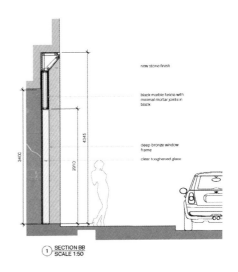

SECTION BB
SCALE 1:50

new stone finish

black marble fascia with minimal mortar joints in black

deep bronze window frame

clear toughened glass

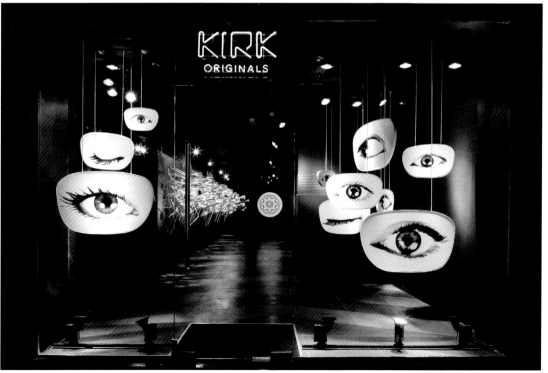

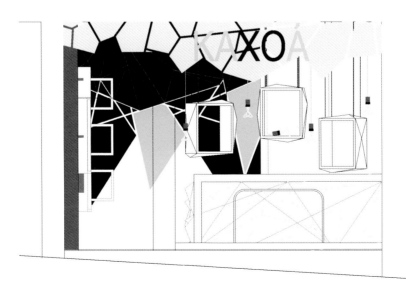

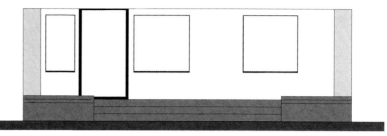
Before renovation
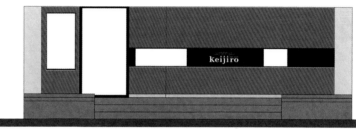

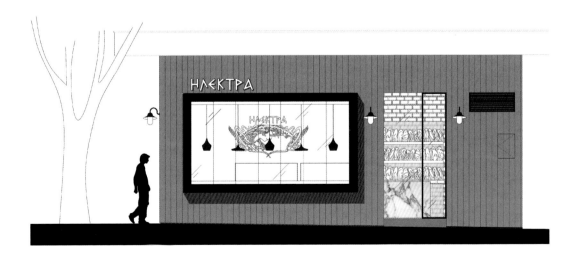

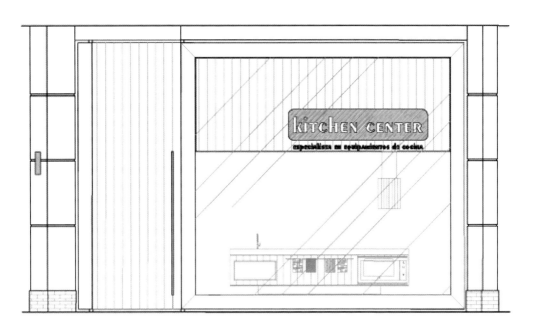

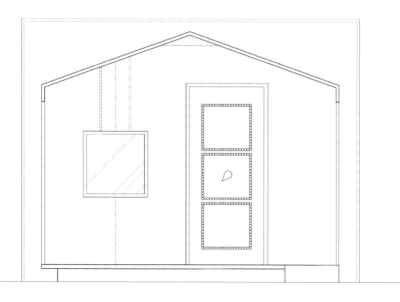

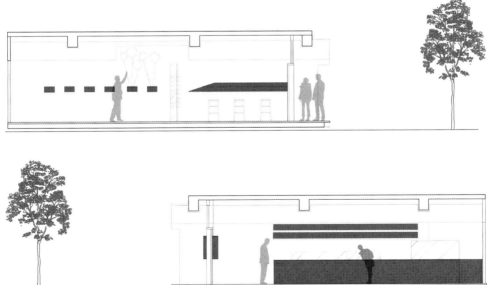

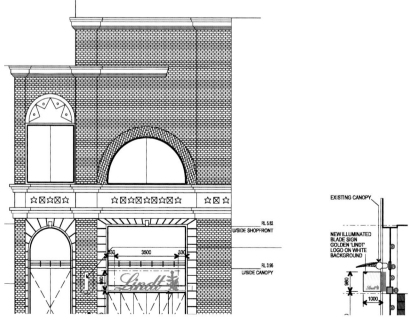

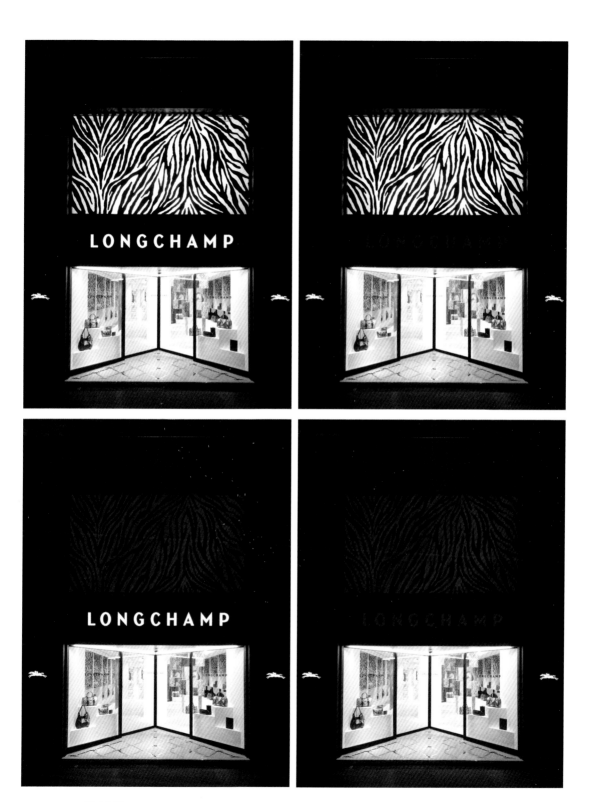

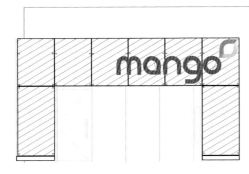

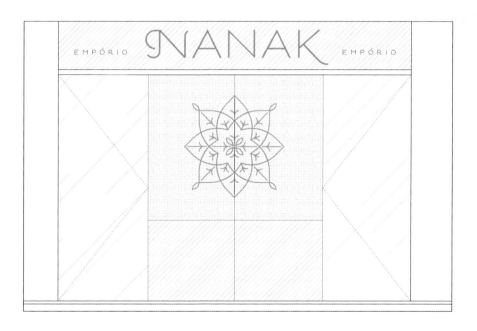

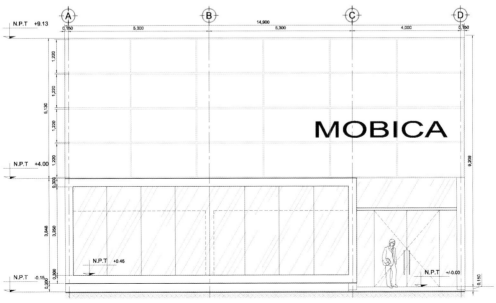

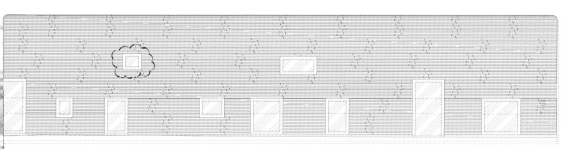

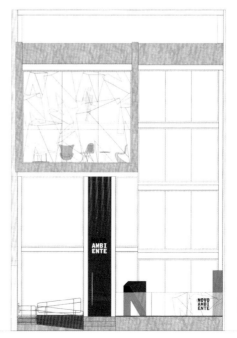

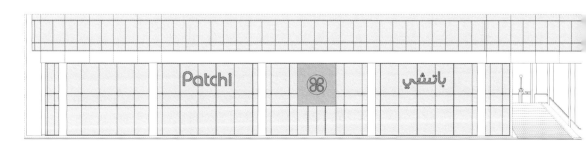

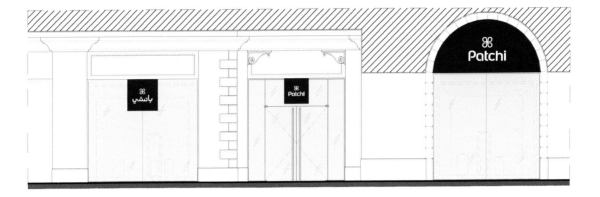

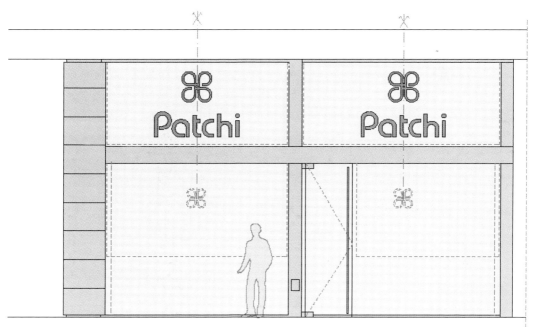

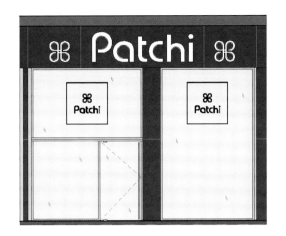

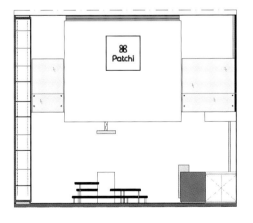
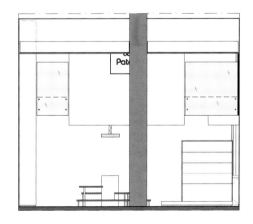

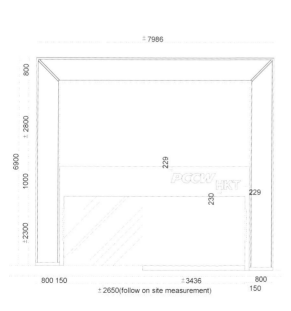

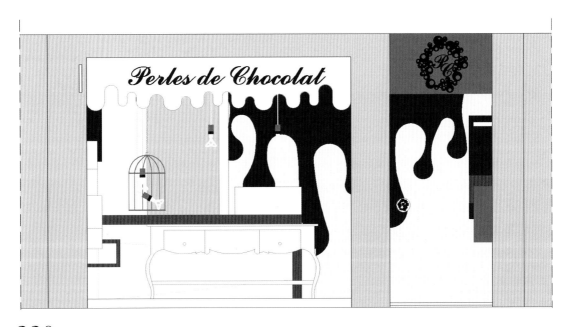

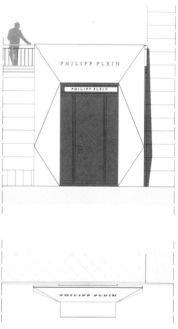

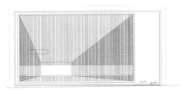
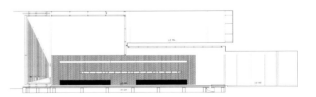

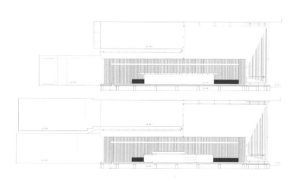

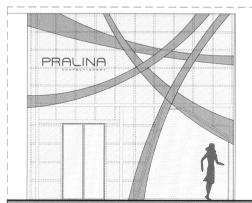

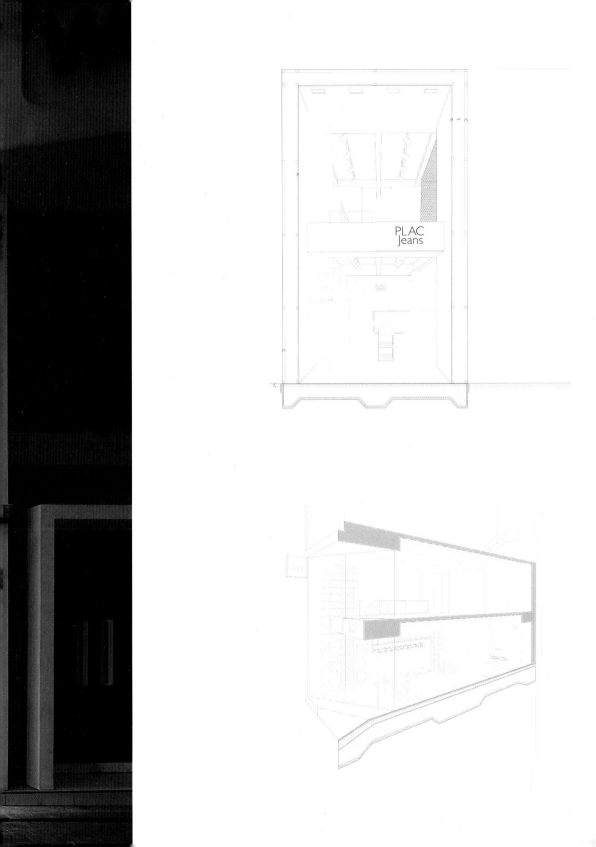

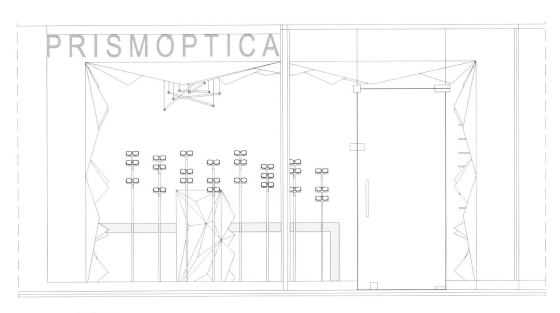

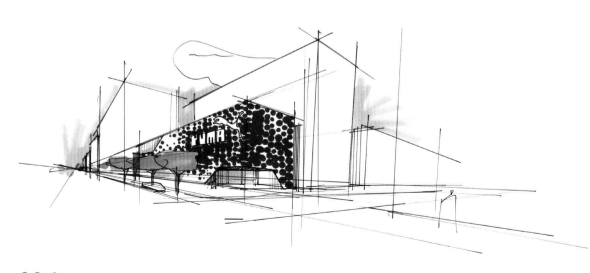

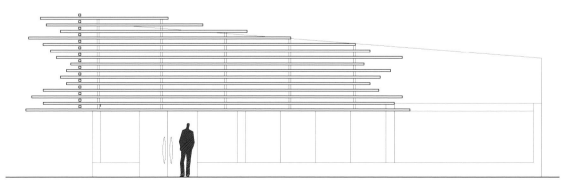

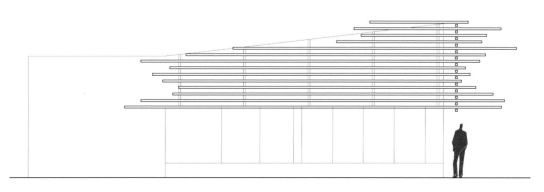

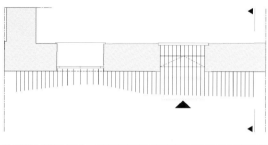

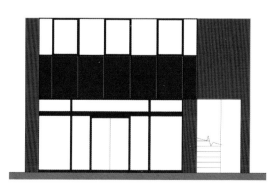

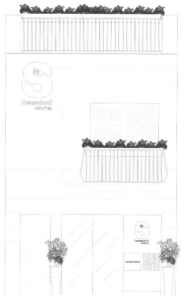

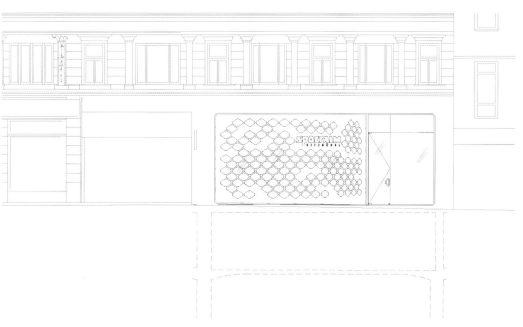

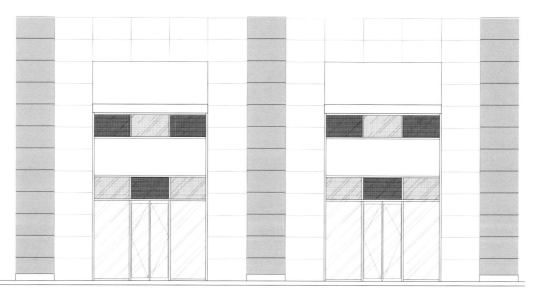

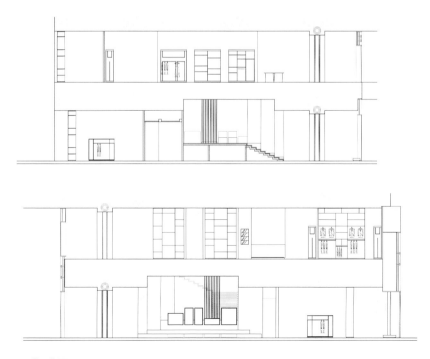

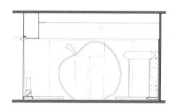
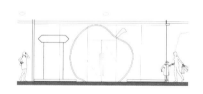
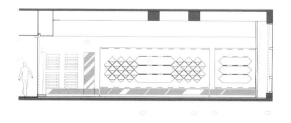
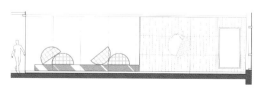

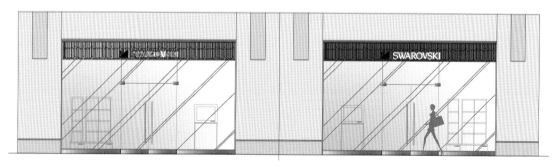

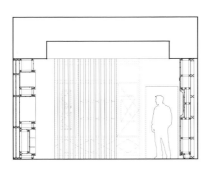
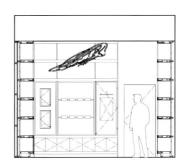
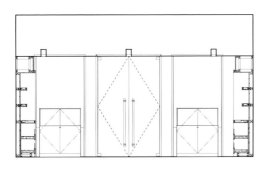
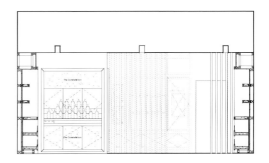

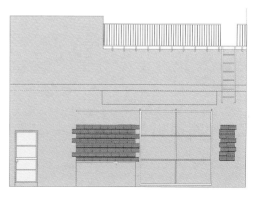

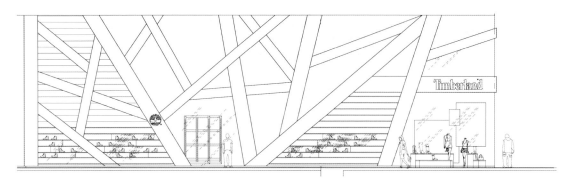

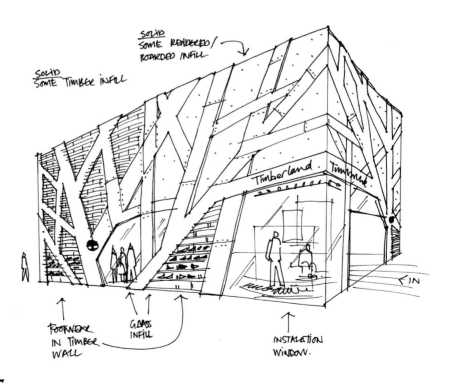

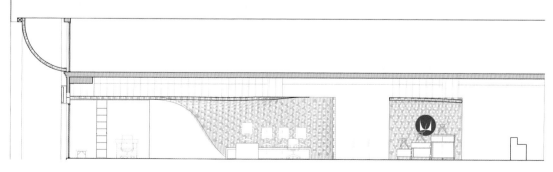

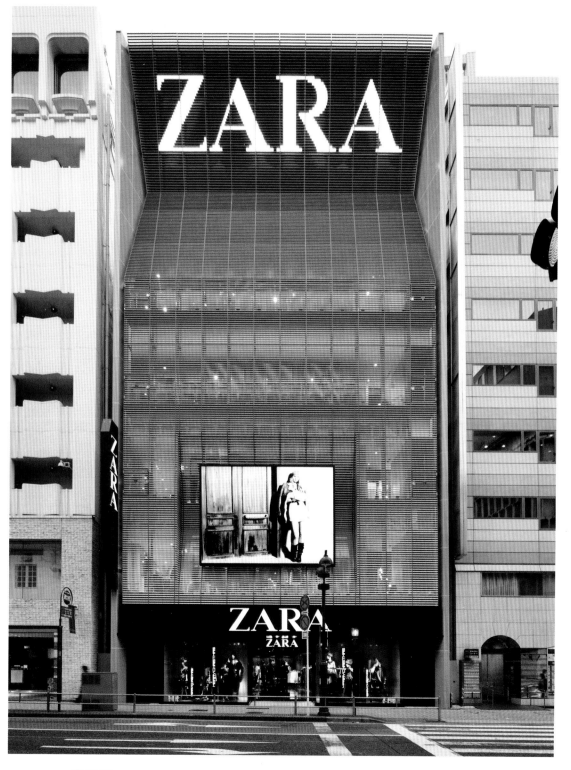

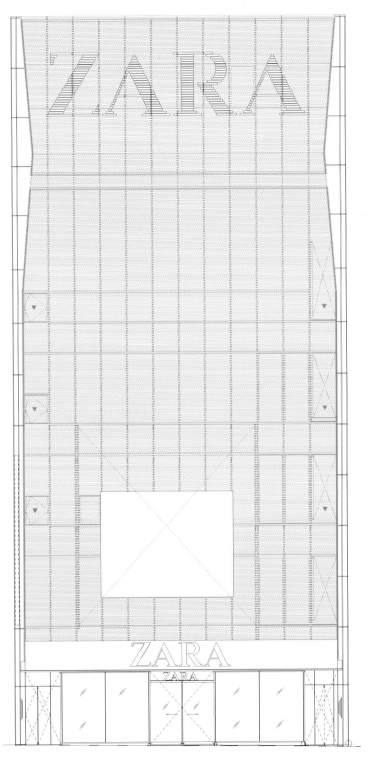

Index
索引

A-cero Joaquin Torres & Rafael Llamazares architects, 106
Ali Alavi, 97
Andres Moore from NODO, 61
Andy Tong Interiors Ltd, 190
AquiliAlberg, 222-223
Architect David Guerra, 138-139
Architect: Keiichi Hayashi, 162-163
Architect\ Castel Veciana Arquitectura, Key Operation, 34-37
ARCO Arquitectura Contemporánea, 188
Arq. Mário Wilson Costa Filho, 154
Arquitectos Interiores, S.C.P., 198-199
Arquitectura en Movimiento, 108-110
arquitectura y diseño, 170-171
Asa Studio + Dom Arquitectura, 262-263
Asprimera Design Studio, 96
Atelier du Pont, 10
BAAR-BAARENFELS ARCHITEKTEN, 264-265
be.bo. - Bel Lobo, Patricia Batista, Alice, 102-103
be.bo. > Bel Lobo, Mariana Travassos, Patricia Fontaine, Patricia Gava, 246-247
be.bo. > Bel Lobo, Patricia Batista, Alice Tepedino, Ana Luiza Neri, Ayla Carvalhaes, Clarisse Palmeira e Fernanda Mota, 227-229
Bercy Chen Studio LP. Calvin Chen, Thomas Bercy, 180-181
Blast Architects, 58-60
Blast Architects, 260-261
Bob Bulcaen /Witblad, 178-179
Bonetti Kozerski Studio, 72-73
Brigada / Damjan Geber, 42-43, 290
Campaign, 150-151
CaoPu, 33
Checkland Kindleysides, 186-187, 284-287
CHIHO&Partners · Kim Chi-ho, 146-149
Chikara Ohno / sinato, 86-87
Christophepillet, 258-259
churtichaga+quadra-salcedo architects, 153
Clifton Leung, 122-123, 216-217

COLLIDANIELARCHITETTO, 76-77
COORDINATION ASIA, 12-13
CuldeSac™, 52-53
Daiko Electric Co.,Ltd, 176
Denys and von Arend, 244-245, 256-257
Droguett A&A Ltda. (DAA), 177
Elia Felices, 11
emmanuelle moureaux architecture + design, 38
Epaminondas Daskalakis, Aggeliki Koutsandrea, 230-231
Estudio Echeverría Edwards, Droguett A&A Ltda, 90-91
Forward Thinking Design, 186-187
Gpstudio, 135, 277
Green Room Design Team, 189
gue y Seta Daniel Pérez + Felipe Araujo, 145
GH+A, 88-89
GWENAEL NICOLAS, CURIOSITY, 175, 288-289
Hitzig Militello Architects, 278-281
Inarch2, 114-115
iks design (Masakazu Kobayashi), 156-157, 254-255
Iosa Ghini Associati – Arch. Massimo Iosa Ghini, 111
Ipanema . Rio de Janeiro, 200-201
Isay Weinfeld, 136-137
JC Architecture, 166-167
Joanna Laajisto Creative Studio, 95
JOSÉ CARLOS CRUZ – ARQUITECTO, 104-105, 107
KAMITOPEN Architecture-Design Office Co.,Ltd., 26-27, 126-127, 144
Keiichiro SAKO, 150, 196-197
Kengo Kuma & Associates, 176
Kiss Miklos, 250-251
Kiyoshi Miyagawa, 54-55
Klab Architecture, 164-165
KNOCK Inc., 134

Koichi Futatsumata, 85
kotaro horiuchi architecture, 142-143
LAURENT DEROO ARCHITECTE, 112-113
lautrefabrique architects, 202-203, 206-207, 208-209, 210-211, 212-215
Luca Nichetto, 274-276
Luciana Carvalho and Renato Diniz, 40-41
MACh Arquitetos, 185
MAM architecture. Pablo Menéndez, 224-226
Manuel García Estudio, 68-69, 272-273
Marcia Arteta Aspinwall & Fausto Castañeda Castagnino, 238-239
Marcia Arteta Aspinwall & Gabriela Kemish & Fausto Castañeda Castagnino, 242-243
MARKETING-JAZZ, 168-169
Masafumi Tashiro, Masafumi Tashiro Design Room, 24-25
Mass Studies, 22-23
Mathias Klotz arquitectos, 184
Matteo Zetti, Eva Parigi, Michela Voglino, 66-67
MIZZI STUDIOS, 252-253
MOBIL M, 16-18, 82-83, 92-93
Monovolume Architecture + Design, 75
MOVEDESIGN, 116-117
MSB architecture and design workshop, 130-131
MVRDV, 78-81
Nendo, 132-133
NicolásLipthay / Kit Corp, 160-161
OFIS ARCHITECTS, 300-301
Omar Dalank, Victor Castro e Carol Kaphan Zullo, 32
P.A.C Pte Ltd, 291
PANORAMA, 20-21, 174
Patrick Kinmonth, Space Architects, 192-195
Plajer & Franz Studio, 236-237
Plajer & Franz Studio Gbr, 159
PLANNING ES (Shinta Egashira), 8-9
PROCESS5 DESIGN, 118-119, 124-125
Ricardo Agraz, 290
ROBERTO FERLITO, 56-57

Rodriguez studio architecture p.c, 84
Rolf Ockert Design, 172-173
ROW Studio, 74
SAKO Architects, 182-183
SAQ-Frederik Vaes, 39
Sergio Mannino Studio, 14-15
Sid Lee Architecture, 94
SO Architecture, 77
Sofia Mora – María Velásquez, 218-219
Sohei Nakanishi Design, 128-129
SOME, 19
Specialnormal Inc, 64-65
SPRS Arquitectura, 155, 220, 234-235
Standard-Jeffrey Allsbrook, 152
Stefano Tordiglione Design, 266-269, 270-271
Stone Designs, 191
Studio Cinque, 62-63
Studio Gascoigne, 101
Studio Marco Piva, 44-47, 48-49, 50-51
Studio mk27, 292-295
Studioprototype Architects, 158, 241
Studio Zara, Key Operation, Matsuda Hirata, 296-299
Studiovase, 140-141
T2.a Architects - Bence TURÁNYI, 71
TAKATO TAMAGAMI ARCHITECTURAL DESIGN, 120-121, 221
The Wall / arquitectura y diseño, 240-241
Unknown – MERCATO, 204-205
URBANTAINER Co., Ltd., 232-233
UXUS, 199, 282-283
Vanja Ilić, 70
Wesley Liu & Eric Lau, 98-100
x architekten, 248-249
Yukio Hashimoto, 28-29
Zouridakis Architects, 30-31

图书在版编目（CIP）数据

店面 / 新空间编辑组编；张晨译.
- 沈阳：辽宁科学技术出版社，2016.3
ISBN 978-7-5381-9552-1

Ⅰ.①店… Ⅱ.①新… ②张… Ⅲ.①商店－室内装饰设计－世界－图集 Ⅳ.①TU247.2-64

中国版本图书馆CIP数据核字(2016)第013608号

出版发行：辽宁科学技术出版社
（地址：沈阳市和平区十一纬路29号 邮编：110003）
印 刷 者：利丰雅高印刷（深圳）有限公司
经 销 者：各地新华书店
幅面尺寸：170mm×225mm
印　　张：19
插　　页：4
出版时间：2016年 3 月第 1 版
印刷时间：2016年 3 月第 1 次印刷
责任编辑：殷　倩
封面设计：周　洁
版式设计：周　洁
责任校对：周　文

书　　号：ISBN 978-7-5381-9552-1
定　　价：88.00元

联系电话：024-23284360
邮购热线：024-23284502
http://www.lnkj.com.cn